INTRODUCTION

The early history of flight in Ireland has been but scantily recorded. This is a great shame for it contains many unique and fascinating aspects that deserve to be brought to a wider audience that those with a particular interest in its story. I was fortunate enough to grow up in an 'aviation' household and from my earliest days the stories of the world's aviation pioneers have held a great fascination for me. In time, as my knowledge of these pioneers increased so my attention turned more and more to the pioneers who first took flight on this island.

My first discovery was that precious little had been written about these pioneers, two notable exceptions being the writings of Kevin Byrne and John S Moore. But my hunger was for more and especially for an understanding of the environment in which these early pioneers existed. Help came from an unexpected source. In my role as curator of the Royal Irish Automobile Club (RIAC) Archive I had access to that wonderful Irish publication, R J Mecredy's *The Motor News*. The RIAC, of course, played its own role in the development of Irish aviation, its members forming the Irish Aero Club in November 1909. In common with other motoring publications of the early 20th century, *The Motor News* carried details of developments in aviation and more particularly developments in Ireland. This material provided a rich source for several of my articles that appeared in *Flying in Ireland* magazine in my column 'Flight Lines'.

These articles aroused some interest and brought forward several suggestions that the appropriate ones might be combined to give a short history of early aviation in Ireland. This Album is the result and it is hoped that it will be the first of several aviation titles from the Dreoilín imprint. I hope the reader will enjoy these tales and will be prompted to learn more about these remarkable men and women who first flew in Irish skies.

Bob Montgomery

For my Father,
John David Montgomery
and all those who have flown the skies
of Ireland

Professor George Francis Fitzgerald experimented unsuccessfully with a Lilienthal glider in the grounds of Trinity College, Dublin, in 1895 and 1896.

EARLY AVIATION IN IRELAND

Bob Montgomery

Dreoilín

CONTENTS

THE DAWN OF FLIGHT IN IRELAND	5
PROFESSOR FITZGERALD'S EXPERIMENTS	6
PROGRESS IN EUROPE AND A BOMBSHELL FROM AMERICA	11
THE FORMATION OF THE IRISH AERO CLUB	14
ENTER HARRY FERGUSON	17
THE REMARKABLE LILIAN BLAND	22
IRELAND'S FIRST AVIATION MEETING	27
YOUNG MASTER McGRATH	32
FIRST ACROSS THE IRISH SEA – ALMOST!	34
THE GREAT DUBLIN TO BELFAST AIR RACE	38
VIVIAN HEWITT AND DENYS CORBETT-WILSON	42
CHRONOLOGY OF FLIGHT IN IRELAND 1784-1912	47

Front Cover: Harry Ferguson airborne over the sands of Newtownards in his revised monoplane. Painting by aviation artist Vincent Killowry.

Published and distributed by Dreoilin Specialist Publications Limited,
Tankardstown, Garristown, County Meath, Ireland.
Telephone: (00353) 1 8354481
e-mail: info@dreoilin.ie

First published in July 2013

Copyright © 2013 Bob Montgomery

ISBN 978-1-902773-29-2

A CIP record is available for this title from the British library.

All rights reserved. No part of this publication may be reproduced or transmitted in any form including photocopy, recording, or any other information storage and retrieval system, without the prior permission from the publishers in writing.

Efforts have been made to trace copyright holders of all photographs/illustrations used in this Album. We apologize for any omissions that may occur, which are unintentional, and would be pleased to include an appropriate acknowledgement in any subsequent edition.

Design by Alan Pepper Design. Set in 9 point Times Ten by Alan Pepper. Design and printed in the Republic of Ireland by Walsh Colour Print.

Partly adapted from the series of 'Flight Lines' articles that first appeared in *'Flying in Ireland'* magazine.

All photographs author's collection unless otherwise stated.

See our full range of books at:
www.dreoilin.ie

On 19th July 1785, Richard Crosbie ascended in his balloon from Leinster Lawn in an attempt to become the first to fly across the Irish Sea.

THE DAWN OF FLIGHT IN IRELAND

On Thursday April 15th in the year 1784, the age of flight dawned in Ireland when a M Rosseau, accompanied by a drummer boy, ascended from a field at Navan, landing an hour and a half later near Ratoath. Contrary to what is generally accepted by historians, this was almost certainly the first flight in the British Isles - a feat more usually attributed to James Tyler who made an ascent from Comely Gardens in Edinburgh in August of the same year.

Rosseau was followed by the remarkable Richard Crosbie who made several sensational flights from Ireland during the years 1785 and 1786 including an attempt to cross the Irish Sea that was to end with his rescue from the sea midway between the coasts of Ireland and Wales. James Sadler, the first Englishman to fly, moved to Dublin to continue his ballooning experiments and in October 1812 he too attempted the perilous crossing of the Irish Sea. Sadler came close to achieving the crossing but was forced to come down in the sea off the coast of the Isle of Man, from where he was rescued.

In the years that followed, the Irish Sea was finally conquered by Windham Sadler - James Sadler's son - who flew from Portobello Barracks in Dublin in 1817 and landed safely near the town of Holyhead in North Wales. Other notable balloonists in Ireland included Edmund Livingstone, John Hampton and John Dunville who with his friend CW Pollock crossed the Irish Sea in 1910, having taken off from the Gas Fields at Ringsend, Dublin, and landed in a snowstorm near Macclesfield in Cheshire.

And there the story of ballooning in Ireland seems to have come to a full stop until the formation of the Dublin Ballooning Club in June 1968, whose members embraced that re-invention of ballooning, the hot-air balloon.

In this photograph Professor Fitzgerald in the Lilienthal glider is being towed in an effort to become airborne.

PROFESSOR FITZGERALD'S EXPERIMENTS

To the German pioneer flyer, Otto Lilienthal, must surely go the credit for the first heavier-than-air aircraft to be produced in any quantity. His most successful glider was his 'No. 11' design of which at least eight are known to have been produced. The British pioneer, Percy Pilcher bought one, as did the Russian, Shukowski, and the American William Randolph Hearst also bought one although there is little evidence that he attempted to fly it. However, there was another 'No. 11' much nearer to home. This was bought in 1895 after some correspondence with Otto Lilienthal, by the famous physicist, Professor George Francis Fitzgerald, of Trinity College, Dublin.

In the course of his correspondence with Lilienthal, Fitzgerald received copious advice on how to assemble the 'kit' as well as how to fly it. The good professor seems to have carefully followed the assembly advice but almost totally ignored the advice on how to fly the 'No. 11'! Having paid Lilienthal £25 for his glider, the machine duly arrived in March 1895. Lilienthal advised Fitzgerald that *"The first trials have to be done very carefully, you have to walk against the wind downhill. When jumping the apparatus must be held horizontal, but when landing it must be upright every time"*.

The professor was not long in making his first attempts at flight. *The Irish Times* of Wednesday April 3rd 1895 reported: *"Shortly after one o'clock yesterday quite a novel exhibition which was witnessed by considerable crowds took place in the college grounds in which a small company had assembled. This was an attempt by*

Professor George Francis Fitzgerald.

A top-hatted Professor Fitzgerald in his Lilienthal No 11 glider in the grounds of Trinity College.

Professor Fitzgerald to float or soar with the aid of an aerial machine that had been specially constructed for the purpose. The machine... is said to be on the model of the latest German invention and is composed of white canvas attached to a strong frame. The machine was gently lowered over the Professor's head and shoulders until they appeared through an aperture in the centre. He then made short runs with the machine which did not rise. Further experiments resulted similarly, but at least one was successful, the machine being raised from the ground, and, aided by a brisk breeze it and the Professor were carried for a short distance."

Despite the detailed advice of Lilienthal, the Professor continued to make his attempts at flying in the grounds of Trinity College where there was no suitable *"gentle slope"* of the type Lilienthal had said was essential. This is hard to understand for there were many suitable sites within easy reach in Dublin's Phoenix Park. Its hard to imagine a more unsuitable site than the grounds of Trinity College where the winds gusted around the many buildings and where there were many trees and the ground itself where the Professor made his attempts was below the level of the surrounding streets.

Fitzgerald had much more success flying the Lilienthal glider as a kite, and these experiments convinced him that *"in a steady breeze, it would be quite feasible to soar with apparatus of this kind."* However, despite further correspondence with Otto Lilienthal, Professor Fitzgerald seems to have lost interest in attempting to fly by early 1896 and made no further experiments. The No. 11 Lilienthal glider was retired and subsequently suspended from the dome of the 'New Building' – later the Museum Building – in the college where apparently a student attempted to set fire to it! It survived this arson attempt and eventually passed into the hands of a Dr McCabe before finally meeting its end at the hands of one of the doctor's friends who attempted to fly from the top of a haystack and *"completely wrecked it"*.

And the Professor? He became a member of the Aeronautical Society of Great Britain in 1897, eventually becoming a council member. Despite his failure to fly with the Lilienthal Glider, he remained fascinated by the possibility of flight until his death in 1901 at the early age of fifty.

The German flight pioneer Otto Lilienthal (1848-1896) was by far the most significant researcher to precede the Wright brothers. Between 1890 and his death in 1896 he made as many as 2,000 gliding flights in a series of machines controlled by shifts in body weight. He constructed an artificial hill near his home in Berlin from which he flew. His scientifically recorded observations were to prove a very significant contribution to the work of those who followed him.

The Brazilian balloonist, Alberto Santos-Dumont, made what is generally agreed to have been the first flight in Europe on 13th September 1906 at the polo grounds in the Bois de Boulogne, Paris, in his 14-bis biplane. His first flight was around 60 yards, and he would return a little over a month later to win the 2,000 franc Ernest Archdeacon prize for the first flight in Europe to exceed a distance of 100 metres. Santos-Dumont was widely feted as the first person to fly, a belief soon to be dispelled by the arrival of the Wrights in Europe.

PROGRESS IN EUROPE AND A BOMBSHELL FROM AMERICA

Would-be Irish aviators were well served by the Irish motoring publication *The Motor News*. This remarkable magazine, founded in 1900 by RJ Mecredy when there were probably fewer than fifty automobiles in the whole of Ireland, began to carry news of aviation experiments from around the world by the middle of the first decade of the new century and by 1908 these occasional items had become a regular aviation column in the magazine.

Thus it was that Irish enthusiasts learned of the success of the Brazilian aviator Alberto Santos-Dumont in making what historians generally agree was the first flight in Europe on September 13th 1906 at the polo ground in the Bois de Boulogne, Paris, in his 14-bis biplane.

A little over a month later, on October 23rd, Dumont followed this by winning the 2,000 franc Ernest Archdeacon Cup when he became the first person in Europe to exceed a distance of 25m in the air with a flight that covered approximetly 60m. Other European would-be aviators soon followed and in March 1907 the Voisin-Delagrange made its first flight piloted by Charles Voisin, followed by Louis Bleriot who made his first flight on April 5th in an aircraft of his own design.

Progress in Europe was now swift and many took to the skies with varying degrees of success. Most notable, perhaps, was Henri Farman, who as part of the French team at the 1903 Gordon Bennett Race in Ireland, had driven a Panhard racing car. Farman soon achieved various records in the air and made the first officially credited circular flight exceeding I km on January 13th 1908, in the process winning the *Grand Prix d'Aviation Deutsch-Archdeacon* prize of 50,000 francs. For a time it must have seemed as if European progress in aviation was unstoppable.

In August 1908, Wilbur Wright stunned French aviators with a series of flights that clearly demonstrated the American brothers superiority in all aspects of aviation over their European rivals. By the time Wilbur returned to make more flights the following year, they were acknowledged as the conquers of the air.

The 1909 flights of Wilbur Wright at Pau attracted many of the crowned heads of Europe as well as large crowds. To these farm workers Wilbur's flights must have seemed miraculous.

Such notions were dispelled when the American Wilbur Wright made his first demonstration flights from the racetrack at Hunaudières, near Le Mans, France, on August 8th 1908. Watched by only a small crowd of curious onlookers, Wright kept the Flyer in the air for just 1 min. 45 secs. at a speed of up to 45 mph and stunned those who saw the flight by making the first banked turns under perfect control ever seen in Europe. As word spread of what had been achieved Wright continued to fly, making nine flights of up to 8 min. duration, the last on August 13th before returning to America.

By September 21st Wilbur was back in Europe and further demonstrated the superiority the two American brothers had achieved with a new European distance record when he flew 41.4 miles in a time of 1 hr. 31 secs. This was by far the longest flight to-date and beat the previous European record flight of 30 min. 27 secs. By year end Wilbur had pushed this record to 2 hr. 20 min. 23 secs., over three times the best flight by a European pilot (Henri Farman).

At year end, Wilbur Wright, now joined by his sister Katherine and brother Orville, set up base at Pau in the south of France. Wilbur made some 60 flights there during January 1909, some of which involved flying lessons for three trainees.

From top left: 1. Otto Lilienthal whose gliding experiments paved the way for the Wright brothers. 2. The Brazilian aviator, Santos-Dumont, who made the first flight in Europe in September 1906. 3. The brothers who conquered flight, Orville and Wilbur Wright. 4. Henri Farman, foremost amongst French fliers.

The founders of the Irish Aero Club photographed in the garage of the Irish Automobile Club on November 5th 1909. Amongst those present were: Johnny Dunlop, second from left, front row; J C Percy, third from left, front row; Harry Ferguson, fourth from left, front row; John Boyd Dunlop, fifth from right, front row; Sir William G D Goff, sixth from right, front row; Edward White, seventh from right, front row. The large circle on the ground is a turntable for turning cars and is still visible in the RIAC garage today. (Photo by courtesy of the RIAC Archive).

THE FORMATION OF THE IRISH AERO CLUB

In Ireland, interest in aviation was growing with experiments being undertaken by several experimenters. Their interest was about to receive a welcome boost. To understand how this came about, one needs to understand the role of the Irish Automobile Club in the pioneering days and development of motoring in Ireland. Founded at a meeting on January 22nd 1901 the Irish Automobile Club (still very much in existence today as the Royal Irish Automobile Club) gave a focus to those adventurous souls who had embraced the new automobilism. In 1901 motoring was still an adventure and setting out on a journey one never knew what calamities awaited or perhaps more importantly, when you might reach your destination!

As a result, early motoring attracted adventurous personalities who enjoyed the freedom offered by the conquest of the road. But by 1907/8 many of the early freedoms were already been curtailed. At the end of 1903, the motoring lobby had conceded the issuing of driving licences and registration marks for cars and motorcycles in return for a 12 mph overall speed limit. (Interestingly, in Ireland through the good offices of the Head of Local Government Sir Henry Robinson, himself a pioneer motorist, a loophole in the legislation was exploited to allow a speed limit of 14 mph). Along with the rapid growth in the number of motor vehicles and the resultant legislation, many of the earliest motorists began to lose interest in 'the great adventure of motoring'.

Meanwhile, all over Europe adventurers and inventors had begun to design and build their own aircraft. Ireland was no different but most were doomed to failure with just a tiny percentage having the ability and perseverance to succeed. Recognising this outbreak of interest the Irish Automobile Club Honorary Secretary Edward White, together with

J C Percy, co-publisher of *The Motor News* weekly magazine, called a meeting for Friday November 5th 1909 at the Club's clubhouse at 34 Dawson Street, Dublin. The Club's Chairman, Sir William G D Goff took the chair and at a well-attended meeting a resolution was proposed by Sir Henry Grattan-Bellew *"that an Irish Club, to be called the Irish Aero Club, be formed, for the encouragement and support of aerial navigation"*. J C Percy seconded the resolution that was passed unanimously.

Edward White then proposed *"that it be referred to the Committee of the Irish Automobile Club to select from the names of those interested in aviation a Committee of not less than ten and no more than twenty to act as the first committee of the Irish Aero Club"*. Mr. Harry Ferguson of Belfast seconded the resolution that also was unanimously adopted. And that, apart from the appointment of Sir William Goff as President and Edward White as Honorary Secretary of the new club, was that.

From these small beginnings things moved quickly forward. Before year end Harry Ferguson had made the first flight in Ireland and in the following year, 1910, again with the assistance of the members of the Irish Automobile Club, the new Irish Aero Club organised Ireland's first Aviation Meeting at Leopardstown followed in 1912 by an air race from Dublin to Belfast. Both of those events are stories we will come to later, sufficient for now to acknowledge the debt Irish Aviation owes for it's birth to the motoring pioneers whose enthusiasm was fired by the exploits of Santos-Dumont and Wilbur Wright and other pioneers.

The Clubhouse of the Royal Irish Automobile Club in Dawson Street, Dublin, photographed in 1910, shortly after the Irish Aero Club was formed by its members at a meeting held here on November 5th 1909. (Photo by courtesy of the RIAC Archive).

ENTER HARRY FERGUSON

The distinction of being the first to fly in Ireland belongs to Harry Ferguson, the remarkable young man from Dromara in County Down. Henry George Ferguson, who was known as 'Harry' throughout his life, was the son of a deeply religious farmer whose ancestors had emigrated from Scotland to the Ulster Plantation in the early seventeenth century.

Harry's formal education ended when he was fourteen and he began work on the family farm. Like another pioneer of machinery, Henry Ford, Ferguson never forgot how hard life on a farm was, and this was eventually to lead him to develop the 'Ferguson System' which was to revolutionize farming and has been called *"the most important de elopment in agriculture in the twentieth century"*. Harry did not like farm work and because of his light build, found it particularly difficult. In 1899, he learned to drive on an early automobile, a Leon Bollée, and began a life-long interest in things mechanical. By 1902, he was about to emigrate to America when his elder brother, Joe, offered him an apprenticeship. Joe had left home in 1895 to become an apprentice maintenance mechanic at Combe-Barbour Mill Engineers. In 1901, Joe, together with another ex-apprentice, set up an engineering business under the name of Hamilton & Ferguson. In 1903 this became JB Ferguson & Company, Automobile Engineers.

Automobile engineering turned out to suit the young Harry Ferguson and he quickly gained a reputation as a fine engine tuner. The company did well and Harry attended night classes at the Municipal Technical Institute. This was to be the only formal technical education he was to have in his life.

The Ferguson monoplane at Donard Park. The 'frames' to support its fuselage during transportation have not yet been removed.

Harry Ferguson, to whom fell the distinction of making the first recorded heavier-than-air flight in Ireland on the last day of 1909 at Lord Downshire's Old Park at Hillsborough. Ferguson also became the first British subject to fly in a machine of his own design and construction.

Ferguson flying over the broad sands of Magilligan during the summer of 1910.

The wreckage of Ferguson's monoplane after his crash at Newtowards in 1911. Ferguson was due to give exhibition flights at the Newtownards Show and was practicing on the previous evening with his mechanic as passenger. On this occasion after landing, the wheels struck a small mud bank, with the result shown here. Ferguson was lucky to escape without injury but his passenger received severe cuts and bruises.

By 1908, Harry, like many of his generation, was excited by the developments that were taking place in Europe and America in the development of aviation. He became interested in the possibility of designing and manufacturing aeroplanes and he traveled to several Air Meetings, including Reims and Blackpool to enable him to study the new flying machines. There, he carefully measured the key dimensions of the various aircraft and on his return to Belfast persuaded his brother Joe that it would be good for their garage business to build and fly aeroplanes.

For his first design Harry adopted the tractor monoplane layout and construction work continued throughout 1909. The Ferguson monoplane had a wing span of 32 feet and a length of 26 feet. It was initially to be powered by a Green engine but Harry decided this would not provide enough power and replaced it with an 8-cylinder JAP engine producing 35 hp. After spending a considerable time experimenting with various propeller designs, Harry successfully flew a short distance on the last day of 1909 at Lord Downshire's Old Park, Hillsborough. In doing so, Harry Ferguson, the young man who had left school at fourteen, became the first person to fly in Ireland and the first British subject to fly in a machine of his own design and construction.

It was an extraordinary achievement by any standards. Ferguson followed up with a flight of a mile at Massarene Park in County Antrim the following April and in June made a number of flights on the wide-open spaces of Magilligan Strand, one of which was over two and a half miles. In August he made a flight of more than three miles to win a prize of £100 and made several flights at Magilligan carrying passengers. In October he suffered a bad crash in which he was knocked unconscious and wrote-off his aeroplane. Undeterred, he rebuilt the monoplane with several major re-design features including a tricycle undercarriage. By June 1911, he had taken to the air successfully again, but in October 1911, he flew for the final time. The reasons Harry Ferguson gave up his flying experiments appear to have been two-fold. Relations with his brother had become strained over the experiments and in 1911 he decided to set up his own garage, May Street Motors, which in 1912 became Harry Ferguson Limited. Quite probably, he felt the need to concentrate on developing this business. It is also likely that he had recognised the high costs involved in developing an aircraft for commercial

The Ferguson monoplane pictured in the difficult weather conditions at Hillsborough Large Park during December 1909.

manufacture.

Harry Ferguson went on to achieve world-wide fame for his inventions in the world of agriculture as well as many automotive developments such as the Ferguson Four Wheel Drive System for cars, but it is doubtful if any of his many achievements were as remarkable as his design and development of the first aeroplane to fly on this island.

Before leaving Harry Ferguson, mention should be made of Joseph Cordner of Londonderry who built a monoplane to his own design in 1908 and whose supporters claim flew before Ferguson's first flight at the end of 1909. One Harry P Swan wrote in 1968 that Joe Cordner had flown his own aircraft at White Strand, Buncrana, in 1908 and had carried a passenger, Frank McGuinness, aloft. Details of the monoplane are scant apart from the two photographs from the *Londonderry Sentinal* newspaper that are reproduced here and show the monoplane at Lough Swilly just before flight trials began. It featured an engine of Japanese make as well as V-shaped openings under the wings and tail that were reported as being of Mr. Cordner's own invention. Unfortunately, the only evidence for Cordner's primacy that has come to light so far is purely circumstantial. Cordner however, continued his interest in aircraft for the rest of his life and received several patents, including one that predates the well-known Handley Page wing slots for improving the lift and handling of aircraft.

First to drive and first to fly in Ireland: John Brown (second left of propeller), the first man to own a car in Ireland and Harry Ferguson (to his right beside propeller), the first Irishman to fly, photographed together with Ferguson's monoplane in the background.

The monoplane of Joseph Cordner of Londonderry whose supporters claim flew at White Strand, Buncrana in 1908. While this cannot be verified, Cordner retained his interest in aviation and went on to receive several aviation related patents.

The Mayfly, designed, constructed and flown by Lilian Bland. This photograph shows it in its early glider form before being fitted with an engine purchased from AV Roe.

THE REMARKABLE LILIAN BLAND

Pioneers in any field tend to be remarkable people, and aviation pioneers perhaps especially so. But amongst those that pioneered flight in Ireland, none is more remarkable that the lady from Carnmoney, Lilian Bland.

The daughter of a well-to-do Belfast family, Lilian was not a typical daughter of her time. As a young woman she wore trousers, smoked cigarettes and was interested in things mechanical. She made a successful career for herself as a photo journalist and was amongst the first, if not the first, to make colour plates of wild birds, on the islands off the west coast of Scotland. She photographed the Irish Gordon Bennett Race of 1903 and also the Airship experiments of Santos Dumont over (and in!) the harbour at Monte Carlo. But it was a postcard of Louis Blériot's monoplane, complete with dimensions, sent from France by her Uncle Robert, which inspired her to want to fly.

The difficulties facing her were enormous. If she wanted to fly, she would have to design and construct a flying machine herself. Assuming she managed to get that far and her design actually flew then she would have to learn to fly it as she flew. Underlying all these difficulties was parental opposition and the considered inappropriateness of a young lady undertaking such a task.

But undertake it Lilian did, and in an utterly professional and scientific manner. She traveled to the first British Aviation Meeting at Blackpool in 1909 and there took detailed notes and dimensions as well as sketches and no doubt, photographs, of all the aircraft there. On her return to Carnmoney she set about teaching herself fluid dynamics so that she could clearly understand what was involved. At the same time she set about designing a model biplane which she flew successfully as a kite. With what she learned in this fashion,

Lilian Bland, the first woman to fly in Ireland.

Lilian then designed and constructed a full-size glider, with the intention right from the start, that if this were flown successfully, she would add an engine at a later date. Constructed of spruce and bamboo with a coating of a photographic solution to render the fabric used to cover the flying surfaces waterproof, the *Mayfly*, as she had christened her 27ft Biplane, was completed early in 1910, just a few months after Harry Ferguson had made the first flight in Ireland on the last day of 1909. Legend has it that when asked why she had named it the *Mayfly,* she replied *"Because it may fly or it may not"*.

Gliding tests were commenced with the help of three men on Carnmoney Hill. These went well and while various breakages showed up several weak spots, these were repaired and strengthened. Lilian next wrote to A V Roe in England enquiring if he could sell her a suitable engine. In reply he wrote to say that he could make for her a two-stroke air-cooled engine – an offer that she accepted. However, the engine was delayed and Lilian traveled to Roe's workshops in July to see a demonstration run of the engine. A horizontally opposed two cylinder engine developing 20hp at 1,000rpm, it's propeller disintegrated on its first test run in the presence of both Lilian and A V Roe, happily without injury to either.

With a new propeller fitted, Lilian took the engine back to Ireland on two bearers in her railway compartment! The engine was soon fitted to the *Mayfly* but proved to produce a severe vibration necessitating further strengthening of the *Mayfly*. Lord O'Neill offered the use of his park at Randalstown for the flight trials. Now ready and eager to fly the *Mayfly*, Lilian and her helpers were forced to sit out five weeks of bad summer weather before the chance to fly finally arose. Soon, Lilian was taking off in about 30ft and making short hops into the air. As experience was gained, *Mayfly* was regularly making longer hops and flying short distances under control. Lilian made several modifications to *Mayfly* as she flew in August and September and over the winter designed and built a test scale model of a proposed 30 ft span evolution of the *Mayfly*.

However, in the meantime, in the face of steadfast opposition to her flying experiments from her father, Lilian agreed to bring them to a stop in return for his offer to buy her a motor car. This car was a Ford which was bought from the Dublin Agent, R W Archer, and soon afterwards, Lilian had obtained the Ford Agency for Ulster. The engine from *Mayfly* was eventually sold and the airframe given away to a Dublin Boy's Club. Sadly, efforts to trace the ultimate fate of this historic aircraft have been in vain.

This remarkable lady continued to live a life of adventure, eventually accepting an offer of marriage from a cousin in Canada. There she lived on virgin land near Vancouver before returning to England in 1935 and retiring to Cornwall in 1955, where she died in 1972.

The Mayfly glider photographed at Carnmoney Hill in 1910.

Lilian Bland at the controls of the Mayfly, here with it's motor fitted.

Flanked by two Huzzars, Drexel's Bleriot monoplane is pulled into position in readiness for flight. (Photo by courtesy of the RIAC Archive).

IRELAND'S FIRST AVIATION MEETING

Earlier we saw how the Irish Aero Club was established in November 1909 at the suggestion of the members of the Irish Automobile Club at Dawson Street, Dublin. The newly formed club quickly moved to further stimulate interest in aviation by staging an Aviation Meeting at Leopardstown racecourse on Monday and Tuesday August 29th and 30th 1910.

The Irish Aero Club secured three notable aviators for the meeting, at which they were to demonstrate their machines – two biplanes and a monoplane. Tremendous interest was aroused in the *"three intrepid birdmen"* as they were invariably described in the various newspapers of the day. Cecil Grace was of Irish-American descent and was a cousin of Sir Valentine Grace who was well-known in Irish legal circles. Grace had been flying for a little over a year and brought a Farman biplane to Ireland. The second of the trio was Captain Bartram Dickson who was an ex-military officer and apparently specialised in gliding with his engine switched off. Like Cecil Grace, he brought a Farman biplane to Ireland. The final member of the trio was also the best-known. J Armstrong Drexel was an American living in Europe and was the holder of several altitude records. Drexel brought two Bleriot monoplanes to Ireland – one of which was capable of taking a passenger aloft.

Special hangers were erected for each of the airmen's machines at Leopardstown and workmen removed the rails from the horse-racing course as well as re-routing telegraph wires. There was tremendous interest in the preparations of the three fliers and great excitement was aroused when Drexel took to the air on the day before the meeting. Drexel had been invited to lunch by Lord Powerscourt and it was his intention to fly to Powerscourt Demesne and to land there. However, on the day, he decided that the weather conditions were too windy and instead traveled to lunch by car.

Cecil Grace's Farman biplane is prepared for flight outside one of the specially constructed hangars during the 1910 Aviation Meeting at Leopardstown. (Photo by courtesy of the RIAC Archive).

Drexel's Bleriot monoplane is readied for flight outside its purpose-built hanger while a large crowd of onlookers await developments. (Photo by courtesy of the RIAC Archive)

August 29th was overcast with frequent heavy rain showers. As the day progressed, however, a wind from the south-east developed and a clearance set-in. The variable weather conditions did not deter the crowds and the *Freeman's Journal* reported that *"by three o'clock the crowd present could only be matched by the Horse Show"*. Amongst the many in attendance were the Lord Lieutenant and his wife, Sir Horace Plunkett and the Secretary of the Aero Club of Great Britain, Mr. H E Perrin.

First to fly was Cecil Grace who flew two circuits of the field before landing. At around 3 pm all of the aircraft were paraded from their hangers to the take-off point and soon after, Grace was airborne again. Drexel flew next, making a total of five circuits of the site and landing after being airborne for a total of eleven minutes. Not so successful was Dixon, whose plane made a short hop of about 300 yards before he brought it to a stop and announced he would need to take it away for an overhaul!

Grace and Drexel flew again. Grace deciding to fly his two-seater and so to a Mr. Desmond Arthur of Ennis, County Clare fell the honour of a footnote in Irish aviation history as the first passenger to fly in Ireland. Drexel, who had already flown out as far as Three Rock Mountain and over Killiney, now flew up to an altitude of around 1,000 feet while Dixon reappeared to fly a circuit of the course. He alleged that someone may have tampered with his machine overnight causing him to change the petrol. With that, the first day's flying at Leopardstown came to an end and the large crowd made its way from the course excitedly discussing the wonders they had seen that day.

Day two was very windy and there was no flying at all until 4 pm when Grace took to the air and made several circuits before landing to loud cheers. Thereafter, Grace confined his flying to short hops and taking passengers up for a charge of £10 per head. Just before 7pm Dickson made one short circuit while Drexel did not fly at all on the second day.

Thus the meeting drew to a conclusion and as the crowd began to stream away from the course there was much looking to the eastern sky, for it was rumoured that a young actor turned aviator named Robert Loraine was making preparations at Holyhead to attempt to become the first to fly the Irish Sea.

The organisers were considerably relived to discover that they had made a profit of £400 from the gate when all expenses were taken into account and there was much speculation that another Aviation Meeting would be held at Leopardstown soon. As things turned out, the Irish Aero Club decided instead to organise an air race from Dublin to Belfast and back in 1912 and choose Leopardstown as the starting point. That event was to lead to tragedy but the first Aviation Meeting at Leopardstown in 1910 can be credited as being the event which first turned Irish eyes skywards and began a record of achievement that continues to this day.

LEOPARDSTOWN 1910

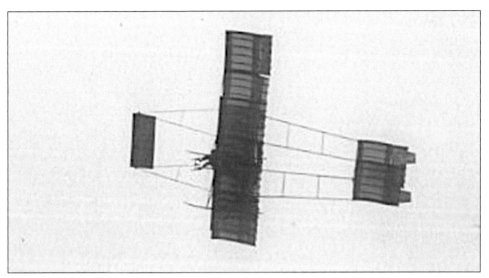

Cecil Grace airborne in his Farman biplane over Leopardstown. (Photo by courtesy of the RIAC Archive).

Drexel in his Bleriot monoplane. Drexel also brought to Leopardstown a two-seater version of the Bleriot in which he took several passengers aloft on the first day of the meeting. (Photo by courtesy of the RIAC Archive).

Grace's Farman finds many helpers as it is pushed back into the cover of one of the several specially constructed hangers. (Photo by courtesy of the RIAC Archive).

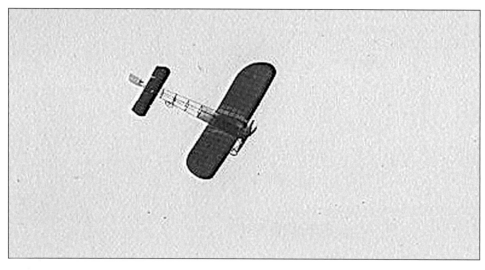

Drexel's Bleriot monoplane in flight. Drexel also flew a two-seater version at Leopardstown. (Photo by courtesy of the RIAC Archive).

Led by two Huzzars, Grace's Farman biplane is wheeled out of its hanger. (Photo by courtesy of the RIAC Archive)

Drexel's Bleriot monoplane in flight receives the rapt attention of the spectators. (Photo by courtesy of the RIAC Archive).

Master J McGrath from Ranelagh flying model aircraft at the Phoenix Parks' Nine Acres in May 1912.

YOUNG MASTER McGRATH

Of all the Aviation disciplines, it is aeromodelling that has the capacity to bestow on the practitioner a deep understanding of how aircraft fly and how they are effected by the natural element in which they fly.

Several years ago I had a discussion with a number of flying instructors during a gliding course at Lasham about the relative merits of learning to fly in a glider versus learning to fly in a powered aircraft. Perhaps not surprisingly, all were agreed that learning to fly in a glider imparted an extra level of ability but one instructor - who has written several textbooks on the subject of gliding – made the point that a pupil who has been a keen aeromodeller brings another level of ability to the task of learning to fly.

I don't doubt this viewpoint for aeromodelling is older than powered flight itself, experimenters flying their creations before the Wright Brothers took to the air in 1903. Here in Ireland, aeromodelling is governed by the Model Aeronautics Council of Ireland and has been practiced for many, many years by a dedicated band of followers, and its highly likely that there were aeromodellers flying their creations here several years before Harry Ferguson became the first to take to Irish skies at the very end of 1909.

Proprietary model aircraft 'kits' were available at least from around 1905 and *The Aero Manual* published by Temple Press in London in 1910 has a most interesting advertisement for 'Gamages – The Premier House for Model Aeroplanes' which reveals how developed model aircraft were at even this early stage. In this advertisement, two petrol engines *"for aero work"* are offered with claimed outputs of 1/4 and 3/4 hp respectively.

Even more interesting is the 'Fleming Williams 27" Racing Monoplane' which it is claimed, is the world's record flyer, having flown a quarter of a mile. The illustration of the Fleming Williams shows a rubber-powered 'A Frame' model driving two contra-rotating propellers. At 63/- it was pretty expensive but no doubt Gamages had less expensive designs available in their catalogue that they promised to send *"Gratis anywhere"* on receipt of a post-card.

Quite who was the first aeromodeller in Ireland is apparently lost to history but we do have clear evidence that the Nine Acres in Dublin's Phoenix park was the flying site of at least one aeromodeller from as early as 1912. (As this site is still in use for model aircraft flying today, does this make it one of, if not the oldest site still in continuous use anywhere?).

The *Motor News* magazine published in Dublin carried a photograph of a Master J McGrath from Ranelagh flying a glider there in it's issue dated May 4th 1912 and commented that the youthful experimenter was to be seen there *"making trials with his gliders"* on Saturday afternoons.

It would be fascinating to know if young Master McGrath developed his interest in aviation and learned to fly a powered aircraft in the years that followed. Indeed, it's not beyond the bounds of possibility that he flew in the terrible conflict that engulfed his generation just two years after the accompanying photograph was taken.

For the present, we'll have to be content with the knowledge that this is the first recorded instance of an aeromodeller in Ireland, and the birth of one of the most fascinating and worthwhile of flying disciplines on this island.

Few photographs exist of the several unsuccessful attempts to build a flying aircraft. This unidentified machine was built in a Donnybrook, Dublin, garden.

Robert Loraine taking off from Rhos-on-Sea prior to his adventurous flight from Holyhead across the Irish Sea.

FIRST ACROSS THE IRISH SEA - ALMOST!

Of all the heroes of early flight who are associated with Ireland, Robert Loraine is one I've always regarded as 'hard done by'. After all, this man was to all intents and purposes the first to cross the body of water that separates us from Britain and what recognition did he get for it? Fame and glory? No, instead he had a cycle club named after him. Hardly seems fair, does it?

Robert Loraine was first and foremost one of the great actors of his day. He became the leading young actor of the London stage – famously making the role of D'Artagnan his own, and then at the height of his fame, he discovered flying and the emerging world of aviation.

It has to be admitted right away that Robert was not a very good pilot and crash followed crash but he was nevertheless determined to pioneer flying as a practical method of transport. Without the Herculean efforts of his French mechanic, Jules Verdrines, who was himself destined to become one of France's most famous aviators, it is probable that he would have achieved little. Nevertheless, by 1910 Robert Loraine had gained sufficient proficiency that he began to plan a very daring exploit – the crossing of the Irish Sea.

At this stage Loraine, who had just ten hours solo flying to his credit, had become known to the public through his flights at the famous Blackpool Aviation Meeting, and after his flight from Blackpool to Anglesey, had left his Framan biplane near Holyhead. Actually, he had crashed it there but it suited his purposes to have the plane at Holyhead without the public being aware of his plans. It would appear that Loraine had planned this flight across the Irish Sea for some time and apparently had no qualms about setting out over the 60 miles of open sea without any naval escort to accompany him and

Robert Loraine, famous actor and aviation pioneer.

with few preparations other than a cork lifebelt and a whistle to blow if he needed to attract attention. His Farman having been repaired and prepared by the faithful Verdrines, he simply started up it's Gnóme engine, took off and climbed to around 4,000 feet before setting off westward and out to sea.

Things quickly began to go wrong, and it became apparent that the Farman had not been properly rigged after it's latest crash. This first became apparent as rigging wires began to snap! No doubt this gave Loraine pause for thought but he soon had another problem to contend with as the engine stopped when he was about twenty miles into the flight. Luckily, a characteristic of the rotary engine is that it turns easily when gliding, and Loraine's began to fire again after he had lost some 2,500 feet in height. Climbing back to 4,000 feet he continued on with his flight. That regaining of altitude was fortunate for soon the Gnóme stopped once more. As he dived in an attempt to restart the engine, more of the rigging wires snapped! This was to happen no less than five times before the Irish coast came into view. Several years later, Loraine was to comment that *"luckily there were plenty of wires on the old Farman"*. Nevertheless, it could only be a matter of time before the process ended in disaster.

By now he could see the coast clearly.

The badge of the Loraine Wheelers Cycle Club took it's inspiration from Robert Loraine's Gnome rotary engine.

His plan had been to land in Dublin's Phoenix Park, but this was quickly abandoned in favour of the closest piece of land. This turned out to be the Bailey Lighthouse at Howth. At first it looked as if he would make landfall beside the lighthouse but as he turned into wind to land, the Gnóme stopped for the last time. As a result, Robert Loraine came down in the sea about seventy yards from the shore. As it happened the members of a newly formed cycle club were enjoying their first outing to the Bailey Lighthouse and watched as the drama unfolded. Several of their members waded out to bring Loraine the last seventy feet to the shore. Thus ended Robert Loraine's epic flight. The Farman was recovered from the sea and sent back to Hendon in several packing cases, where, on this occasion, it was not repaired. The cycle club, who were without a name, decided shortly afterwards to call themselves The Loraine Wheelers Cycle Club, recalling the strange adventure that had met them on their inaugural outing. As for Robert Loraine, he continued to fly, crashing often, but between times conducting the first experiments with air to ground wireless, something that was to prove very valuable in the coming conflict. In the Great War, Robert joined the Royal Flying Corps and became one of it's most controversial commanders helping to pioneer night flying, before returning to his career in the theatre after the end of hostilities.

Astley starting from Leopardstown. (Photo by courtesy of the RIAC Archive)

Police guard Valentine's plane at Leopardstown. (Photo by courtesy of the RIAC Archive)

THE GREAT DUBLIN TO BELFAST AIR RACE

In September 1912, the Aero Club of Ireland, no doubt stung by criticism that it had done little since its foundation in November 1909, to further aviation in Ireland other than organize the first Aviation Meeting at Leopardstown in 1910, decided to hold an air race from Dublin to Belfast and back again.

Once again, Leopardstown was the site chosen for the start of the race and on a September Saturday morning (7th) a large crowd assembled at an early hour to see the aviators off. That considerable advances had been made in aviation since the 1910 Aviation Meeting at Leopardstown, was evident from the variety of aircraft and flyers present for the race. HJD Astley brought his well-used Bleriot monoplane, Lieut. Porte his Deperdussin, Desmond Arthur a Bristol and James Valentine was entered in another Deperdussin. Also present was the well-known flyer, M. Salmet,

Clockwise from top left:

1, Major Wellesley announces that Salmet will give an exhibition flight. 2, James Valentine (with back to the camera) is seen with Mr. R Perrin, the secretary of the Royal Aero Club and HJD Astley. 3, Desmond Arthur and the chairman of the Aero Club of Ireland, Mr. J Dunville at Leopardstown. 4, Lord and Lady Powerscourt in discussion with Mr. Delacombe, Astley's manager, prior to the start. (All photos by courtesy of the RIAC Archive)

Desmond Arthur in the cockpit of his Bristol. (Photo by courtesy of the RIAC Archive)

HJD Astley makes ready his Bleriot monoplane prior to the start of the race. (Photo by courtesy of the RIAC Archive)

J Valentine, M. Henri Salmet and HJD Astley. *(Photo by courtesy of the RIAC Archive)*

Salmet about to take off from the confined space available at Balmoral Show Grounds. *(Photo by courtesy of the RIAC Archive)*

Astley, who was tragically killed at Balmoral, with his wife at Hendon. (Photo by courtesy of the RIAC Archive)

Salmet flying into the grounds at Balmoral. (Photo by courtesy of the RIAC Archive)

HJD Astley. (Photo by courtesy of the RIAC Archive)

who had been engaged by the *Daily Mail* to give an exhibition *"of spectacular flying"*.

All morning the wind blew very strongly, gusting from different directions at times. At 1.00 pm a telephone call came from Harry Ferguson in Belfast to say that the weather conditions were very bad there with high winds and a drizzle that cut visibility right down. In Leopardstown the weather improved a little and M. Salmet took to the air in an attempt to give his promised exhibition. In truth, his flight turned into a quite different exhibition than he had intended, as it was immediately apparent that regaining the ground without mishap would be a major achievement. This, however, Salmet managed to do after an eight-minute flight, receiving the congratulations of his fellow flyers for his skill in doing so.

As the afternoon was now wearing on, it was decided that if the race should start it would end at Belfast as there would be insufficient time for the return flight before darkness. At 4.00 pm it was reported that the weather in Belfast had considerably improved and a large crowd had gathered at the finish point at Balmoral Showgrounds in the expectation of seeing the aviators. Ten minutes later, Astley took off followed a few minutes later by Valentine. Desmond Arthur was next to go but not before he had cleared a bad misfire in the engine of his Bristol. However, his machine refused to lift and realizing he was running out of runway he turned to the right to avoid running into the crowd, in the process bursting a tyre and hitting a small flagpole with one of his wingtips close to the Press Tent.

By now the wind had increased once more as Lieut. Porte found to his cost as soon as he became airborne. After a series of 'swoops' as he recovered from particularly violent gusts, he turned at Stillorgan and returned to Leopardstown and declared conditions too bad to fly. Desmond Arthur, meanwhile had discovered the flagpole had

J Valentine, who set a record for distance by flying from Leopardstown to Newry. (Photo by courtesy of the RIAC Archive)

damaged his wing more than at first realized and so he was out of the race before he had started.

The two remaining flyers, Astley and Valentine, continued. Astley flew higher and encountered even stronger winds, while Valentine made better progress at a lower altitude, overhauling his rival before the wind became even stronger. Astley decided to turn back to Leopardstown but was forced to come down near Drogheda. Valentine got as far as Newry, before he too was forced to descend, landing with great difficulty in a small field two miles from the town. His one consolation was that he had established a record for the longest flight yet achieved in Ireland.

So the great race ended. The first prize of £300 was divided equally between Astley and Valentine, the 'Shell' prize of £50 was awarded to Lieut. Porte and the sum of £25 to Desmond Arthur. In addition each flyer received £40 expenses. Aware that the large crowd at Balmoral had been robbed of their chance to see the aircraft, Astley and Valentine undertook to give demonstration flights there on the following Saturday. M. Henri Salmet also decided to fly there - which he did without incident - and entertain the crowd. Several short flights and another impressive display by Salmet entertained the crowd and such was their response that Astley and Valentine arranged to repeat the exhibition on the following Saturday. On the appointed day, Valentine flew first and then Astley took to the air. Still low, he made frantic signals to the crowd ahead of his flight path, as he suddenly appeared to be in difficulties. The crowd neither saw or understood his signals to scatter and Astley banked sharply to the right in an effort to avoid them. This resulted in his Bleriot falling to the ground, Astley dying almost immediately. At the subsequent inquest, the jury added a special rider to its verdict that *"the aviator, realizing the situation, chose rather to sacrifice his life than carry destruction into the midst of the large crowd."*

Following the air race, Valentine's Deperdussin monoplane was towed by a Ford 20hp car to Mullingar, Cavan, Carrick-on-Shannopn, Castlebar and Tuam at all of which locations demonstration flights were made. (Photo by courtesy of the RIAC Archive)

Corbett-Wilson's flight ended in a furze-covered bank at Crane, close to Enniscorthy.

VIVIAN HEWITT and DENYS CORBETT-WILSON

The next attempt came in April 1912 when two friends, Denys Corbett-Wilson from Kilkenny and Damer Leslie Allen from Limerick, both pupils at the Bleriot Flying School at Hendon, decided to attempt a flight in company to Dublin. Each flying a Bleriot monoplane, they flew in company from Hendon but soon became separated, Allen taking a much more northerly course. After stops near Hereford, Chester and Crewe, Allen took off again and was spotted heading out to sea near Holyhead. Sadly, this was the last anyone was to see of Allen who had foregone the addition of a lifejacket to his equipment in an effort to save weight.

Meanwhile, unaware of this unfolding drama surrounding his friend, Corbett-Wilson took a more southerly course to finally arrive at Fishguard. On the morning of his sixth day (April 22nd) out from Hendon he left Fishguard and flew across the St. George's Channel in a time of 1 hour 40 minutes, landing at Crane, close to Enniscorthy. His flight was not without its own drama as his engine gave trouble throughout most of the flight and when he was about 20 miles from the Wexford coast, he flew into a severe rain-storm which had the effect of reducing his visibility to nil. When he finally came out of this rain storm it was to see the welcome sight of green fields below him. Even then his troubles were not at an end for attempting to land in a large field, his Bleriot's momentum carried him on into a furze-covered bank, the impact breaking his propeller and under-carriage.

Denys Corbett-Wilson

Corbett-Wilson had become the first man to fly from Britain and make landfall on Ireland but even as he was being acclaimed for his daring feat, Englishman Vivian Hewitt was about to set out to fly from Anglesey to Dublin. Hewitt, like Allen and Corbett-Wilson, was also piloting a Bleriot monoplane. However, strong gales kept him grounded at Anglesey for four days and it was not until April 26th that he took off from Holyhead and set course for Dublin.

By all accounts, unlike his fellow pilots flights between the two islands, Hewitt's flight was trouble-free and after a journey of 1 hour 15 minutes he landed at his intended destination in Dublin's Phoenix Park.

Technically, Hewitt was the only one of the four to have crossed the Irish Sea, Corbett-Wilson's flight having been across the St. George's Channel and Robert Loraine having fallen short in his effort. I rather think it's churlish to deny Robert Loraine the honour of being first across the Irish Sea even though the

Hewitt ended his cross channel flight in Dublin's Phoenix Park, where his Bleriot monoplane was soon surrounded by a curious crowd.

honour is usually given to Corbett-Wilson who did not, in fact, actually fly across the Irish Sea. That bible of early aviation first published in 1933, *History of British Aviation 1908-1914* by R Dallas Brett credits Robert Loraine as being the first to fly the Irish Sea, a judgment I'm happy to agree with.

The Welsh aviator, Vivian Hewitt, who successfully flew from Holyhead to Dublin.

Hewitt receiving telegrams of congratulation at the entrance to the Hibernian Military School, where he rested after his flight.

CHRONOLOGY OF FLIGHT IN IRELAND
1784 - 1912

1784 - April	Rosseau ascends from a field at Navan – the first flight in the British Isles.
1786	Richard Crosbie makes failed attempt to cross the Irish Sea by balloon.
1812 – October	James Sadler fails in his attempt to cross the Irish Sea.
1817	Windham Sadler successfully crosses the Irish Sea flying from Portabello Barracks in Dublin to near Holyhead in North Wales.
1895 - April	Professor GF Fitzgerald begins his experiments with a Lilienthal No.11 glider.
1908	Supporters of Joseph Cordner claim he successfully flew his monoplane at White Strand, Buncrana. However, evidence for his primacy is purely circumstantial.
1909	The daughter of a well-to-do Belfast family, Lilian Bland, travels to the first British Aviation Meeting at Blackpool, there taking detailed notes and dimensions as well as sketches of all the aircraft.
1909 – November	The Irish Aero Club is formed at a meeting in the Irish Automobile Club Clubhouse in Dawson Street, Dublin. Sir GWD Goff is elected first chairman.
1909 – December 31st	Harry Ferguson successfully flies his monoplane at Lord Downshire's Old Park, Hillsborough, becoming the first person to fly in Ireland and also the first British subject to fly in a machine of his own design and construction.
1910	Early in the new year, Lilian Bland commences flying experiments with her Mayfly glider on Carnmoney Hill.
1910 – August	Ferguson wins a prize of £100 for a flight of more than three miles.
1910 – August	Lilian Bland makes several 'hops' in the Mayfly, now powered by an engine supplied by AV Roe.

1910 – August	The Irish Aero Club organises Ireland's first Aviation Meeting at Leopardstown horse racing course, Dublin.
1910 – September	Lilian Bland is regularly making flights of a quarter mile in the Mayfly.
1910	Robert Loraine falls short of crossing the Irish Sea by just several hundred yards. He is rescued from the sea just off the Bailey Lighthouse, Dublin.
1910 – October	Ferguson suffers a bad crash, being knocked unconsious and wrecking his monoplane.
1911 – June	Ferguson takes to the air again in his rebuilt monoplane with several modifications including a tricycle undercarriage.
1911 – October	Ferguson flies for the last time and ends his flying experiments.
1912 – April	Damer Leslie Allen attempts to fly the Irish Sea and is lost in the attempt.
	Allen's friend, Denys Corbett-Wilson crosses the St. George's Channel from Fishguard to Crane, near Enniscorthy.
	Vivian Hewitt flies from Anglesey to Dublin's Phoenix Park, becoming the undisputed first to cross the Irish Sea by aeroplane.
1912 – September	The Aero Club of Ireland organises an air race from Dublin (Leopardstown) to Belfast (Balmoral Showgrounds) and back to Dublin. Four aviators start but all are forced down by high winds. Because of the large crowd that had gathered to see the flyers at Balmoral, it is decided to give several 'demonstration' flights there on the following Saturday. On one of these Astley crashes heavily and is killed, the first fatality in Irish aviation.